STOP TO NUCLEAR

Dreaming of a World Without Nuclear Energy and Atomic Bombs

John Valentine

Cod ISBN: 9798861491860

Cover design by: Art Painter
Library of Congress Control Number: 2018675309
Printed in the United States of America

To those who dream of a safer and more peaceful world, this book is dedicated to you. To the curious, the activists, the students, and anyone who wishes to better understand the complex implications of nuclear energy and atomic weapons.

It is for those who believe that knowledge is the light that dispels the darkness of ignorance, that diplomacy can overcome hostility, and that art and culture have the power to inspire change. This book is for those who strive to make a difference in a world often dominated by fear and division.

To the survivors of nuclear accidents, to the victims of the horrors of nuclear weapons, and to all those who have suffered because of the nuclear, your resilience and courage are an endless source of inspiration.

To the young people who will inherit the future of our planet, who are filled with ideas and ideals, we ask you to take up the torch and carry on the fight for a nuclear-free world.

This book is a call to action, an invitation to reflect, and an opportunity to learn. May it illuminate your path and strengthen your commitment to a world where peace prevails over the nuclear threat.

With hope and determination,

John Valentine

There is extraordinary power in our ability to imagine a world without nuclear weapons. It is in that dream that we find the strength to transform reality.

JOHN VALENTINE

CONTENTS

INTRODUCTION

Welcome to "Stop the Nuclear: Dreaming of a World Without Nuclear Energy and Atomic Bombs." Let's begin this journey by exploring a complex world, dominated by forces as powerful as they are ambivalent. Nuclear, whether in the realm of energy or weapons, has shaped our planet in deep and irreversible ways. It is a force that has the potential to generate prosperity or destroy life on Earth.

Nuclear is a topic that cannot be ignored. It represents a scientific, political, and ethical challenge. This book never takes sides but rather seeks to provide a balanced overview of the various dimensions of nuclear, from its role in power plants to its threat as a tool of mass destruction.

In the following pages, we will explore the origins of nuclear energy and its early development. We will discover the catastrophic potential of nuclear weapons and the lessons learned from nuclear incidents like Fukushima and Chernobyl. We will also examine the hidden costs of nuclear energy, both in environmental and health terms, and analyze the influence of the nuclear industry on politics.

We will highlight the crucial role of international organizations in nuclear control and take a look at global efforts towards nuclear disarmament. We will explore how science and technology can contribute to promoting disarmament, and we will analyze the role of charismatic leaders in this process.

We will also examine the landscape of peace and nuclear disarmament movements worldwide, giving voice to nuclear victims and reflecting on how art and culture can raise public awareness. We will imagine a world without nuclear weapons and consider the geopolitical tensions that hinder disarmament.

Finally, we will explore the role of education, non-governmental organizations, and individual responsibility in promoting nuclear disarmament. And we will provide a roadmap for a nuclear-free world, outlining concrete steps to achieve this goal.

This book is an invitation to reflection, an exploration of one of the most urgent and complex challenges of our time. It is a call to action, as nuclear is a topic that requires the participation of all of us. Whether you are scholars, activists, politicians, or curious citizens, we invite you to explore the world of nuclear with an open mind and a critical spirit.

We are only at the beginning of this journey, and the fate of nuclear and nuclear disarmament depends on the choices we make. We invite you to join us as we examine this crucial topic, in search of solutions and hope for a safer and more peaceful future.

Safe travels,

John Valentine.

PREFACE

Welcome to this literary adventure, on a journey through the deep and complex waters of nuclear, where energy and weapons intertwine in a story of scientific progress and global challenges. I invite you to explore the pages of "Stop the Nuclear: Dreaming of a World Without Nuclear Energy and Atomic Bombs."

This book was born from the belief that nuclear is one of the most urgent and significant themes of our time. From nuclear energy to atomic weapons, nuclear has shaped our world in ways that often elude immediate understanding. It is a powerful force that can be both an ally and a threat to humanity, depending on how it is managed.

Nuclear energy has promised a source of clean and abundant electricity but has also generated radioactive waste and fueled fears of atomic disasters. Nuclear weapons, meanwhile, have kept the world in an unstable balance during the Cold War and beyond but continue to hang like a sword of Damocles over humanity.

Throughout these pages, we will delve into the origins of nuclear energy, explore the tragedies of Fukushima and Chernobyl, and analyze the catastrophic potential of nuclear weapons. We will discover the role of international organizations and global movements in promoting nuclear disarmament and take a glimpse into the uncertain future of this crucial field.

"Stop the Nuclear" is a call to action, an appeal for greater awareness, and an invitation to dream of a world without nuclear. It is a work that seeks to provide in-depth information while also encouraging critical reflection and open discussion.

Nuclear energy and atomic weapons are not just matters of politics or science but are issues that involve the fate of humanity and our planet. As individuals, as communities, and as nations, we have the responsibility to address these challenges and work together for a safer and more peaceful future.

This book has been written with the hope that we can find the courage to confront the complexity of nuclear, the willingness to promote peace, and the determination to say "Stop the Nuclear." Every page is an invitation to join this important dialogue, to contribute ideas and actions, and to dream with us of a better world.

Happy reading, and thank you for being part of this quest for nuclear peace.

With gratitude and commitment,

John Valentine.

PROLOGUE

In a hidden corner of the world, where science meets politics, where the promise of progress clashes with the shadow of destruction, the story of nuclear unfolds. It's a story that winds through research laboratories and diplomatic meeting rooms, amidst towering nuclear power plants and atomic weapons hidden in underground silos. It's a story that has altered the course of human history, creating extraordinary opportunities and imposing unimaginable threats.

The idea of splitting the atom, that tiny particle forming the basis of matter, was one of the most revolutionary scientific discoveries of the 20th century. The ability to harness the energy trapped within atoms opened up new frontiers for humanity. Nuclear energy pledged an inexhaustible resource to power our ever-growing need for electricity. It promised a solution to the world's energy and environmental challenges.

However, with this promise came a terrifying menace. The very science that made nuclear energy possible also made the creation of atomic weapons of unprecedented power possible. The bombings of Hiroshima and Nagasaki in 1945 marked a turning point in human history. Those explosions shook the world and heralded the beginning of the nuclear age.

Nuclear weapons brought with them the potential for global-scale destruction. World leaders realized that possessing nuclear weapons would secure them a seat at the table of global decision-making but would also hang a sword of Damocles over humanity. The Cold War was a period when the world was divided between two superpowers armed to the teeth, ready to pull the nuclear trigger at any moment.

Amidst this instability, nuclear energy grew as a crucial energy resource. Nuclear power plants were constructed worldwide, promoting the promise of a continuous supply of clean electricity. But with this development came a new set of problems: the management of radioactive waste, the risk of nuclear accidents, and the constant concern about nuclear proliferation.

Over the decades, nuclear policy has become a complex battleground where government decisions can have global impacts. International agreements were signed to limit the spread of nuclear weapons, but the challenge of disarmament remains elusive. While some nations have dismantled their nuclear warheads, others have sought to develop new ones.

Yet, there is hope. Throughout history, the world has seen charismatic leaders actively pursuing nuclear disarmament. International organizations such as the International Atomic Energy Agency (IAEA) and the Treaty on the Non-Proliferation of Nuclear Weapons (NPT) have worked to maintain nuclear order.

Movements for peace and nuclear disarmament have gained momentum worldwide. Ordinary people have joined forces to advocate for a world without nuclear weapons, engage in awareness campaigns, and influence their governments' policies. Art and culture have played a significant role in promoting this cause, inspiring millions to contemplate nuclear issues in new and creative ways.

This book is an attempt to explore nuclear in all its facets. From science to politics, from history to dreams for the future, we are prepared to dive into this complex world. We will uncover the origins of nuclear energy, explore the devastating effects of atomic weapons, and analyze the hidden costs of nuclear energy. We will examine the role of international organizations in nuclear control and endeavor to understand the contribution of science and technology to nuclear disarmament.

Moreover, we will engage with the stories of charismatic leaders who have advocated for nuclear disarmament and explore global movements for peace and nuclear disarmament. We will give voice to nuclear victims and seek to comprehend how art and culture can raise awareness and inspire public opinion.

This book is also a call to action. It is an invitation to dream of a world without nuclear and to work together to realize that dream. Nuclear can be an ally or a foe, but its direction depends on the choices we make. We are the architects of our future, and nuclear is one of the most pressing challenges we must face.

Are you ready for this journey? Prepare to explore nuclear in all its dimensions, reflect on our responsibility as individuals and as a society, and dream of a world without nuclear.

CHAPTER 1:
THE ORIGINS OF NUCLEAR ENERGY: A HISTORICAL ANALYSIS OF EARLY APPLICATIONS

Nuclear energy is a concept deeply rooted in human history, although practical applications only date back to the 20th century. The discovery and development of this form of energy have shaped our world in profound and complex ways, bringing with it both extraordinary benefits and unsettling dangers. In this chapter, we will explore the origins of nuclear energy and the early applications that laid the groundwork for its evolution.

The Discovery of the Atomic Nucleus and J. Robert Oppenheimer's Contribution

The history of nuclear energy begins with the discovery of the atomic nucleus. In 1911, physicist Ernest Rutherford conducted the alpha scattering experiment, demonstrating that atoms contained a dense, positively charged nucleus at the center, surrounded by orbiting electrons. This discovery laid the foundation for understanding the atom and paved the way for future developments in nuclear physics.

Another crucial step in understanding nuclear energy was taken by Albert Einstein in 1905 when he formulated the famous equation $E=mc^2$ in his theory of special relativity. This equation revealed the profound connection between mass and energy, suggesting that a small amount of matter could be converted into a large amount of energy. This concept would be fundamental to the development of future nuclear technologies.

However, we cannot discuss the origins of nuclear energy without mentioning the significant contribution of J. Robert Oppenheimer. In the 1940s, Oppenheimer led the Manhattan Project, a secret research program that led to the development of the first atomic bombs. His leadership in this historic project underscores the crucial role of nuclear research in the military context during World War II and marks a significant milestone in the history of nuclear energy.

Marie Curie and the Discovery of Radium

In 1898, Marie Curie and her husband Pierre Curie discovered radium, a highly radioactive element. This discovery not only contributed to the understanding of nuclear instability but also opened the possibility of harnessing the intrinsic energy of nuclear reactions. However, at the time, radioactivity was still poorly understood, and its health hazards were underestimated.

The First Controlled Nuclear Reaction

On December 2, 1942, under the bleachers of a football stadium at Chicago Pile-1, Italian American scientist Enrico Fermi achieved the first controlled nuclear reaction in human history. This event marked a fundamental step toward the practical application of nuclear energy. Chicago Pile-1, a rudimentary nuclear reactor, demonstrated that it was possible to control a nuclear reaction, producing energy safely and under control.

The Atomic Age and the Cold War

The end of World War II marked the beginning of the Atomic

Age, with the dropping of the first atomic bombs on Hiroshima and Nagasaki in 1945. These events showcased the destructive power of nuclear weapons and marked the onset of the Cold War between
the United States and the Soviet Union. The nuclear arms race became a significant aspect of international relations, fueling research and development of atomic weapons.

Nuclear Energy for Civilian Use

In parallel with the development of atomic weapons, the idea of using nuclear energy for civilian purposes emerged. In 1954, the Soviet Union launched the first nuclear reactor for electricity production, marking the beginning of the nuclear era for peaceful purposes. This initiative generated enthusiasm for nuclear energy as a promising source of low-cost electricity.

The Chernobyl Disaster and Its Consequences

Despite the benefits of nuclear energy, the world had to face the devastating consequences of the Chernobyl accident in 1986. This nuclear disaster revealed the inherent risks associated with nuclear energy and fueled the debate about its safety and sustainability. Chernobyl demonstrated that, even though nuclear reactions can be controlled, human error or malfunction can have catastrophic consequences.

Conclusions

The origins of nuclear energy are complex and multifaceted, with significant contributions from scientists, physicists, and engineers worldwide. This historical analysis has shown how the discovery of the atomic nucleus, Einstein's theory of relativity, Marie Curie's pioneering research, and the crucial work of J. Robert Oppenheimer laid the foundations for nuclear energy as both a source of progress and potential destruction. Throughout the 20th century, nuclear energy has become a fundamental part of our lives, serving as both a source of power and a double-edged sword. In this book, we will explore how our understanding of nuclear energy has evolved over time and how we can address the challenges it presents for the future of our planet.

CHAPTER 2: THE CATASTROPHIC POTENTIAL OF NUCLEAR WEAPONS

Nuclear weapons represent one of humanity's greatest technological achievements, but also one of the gravest threats to life on Earth. In this chapter, we will explore the catastrophic potential of nuclear weapons, including their devastating effects, both immediate and long-term, on the planet and human civilization.

The Nuclear Inferno: Immediate Effects

The most common image associated with nuclear weapons is that of destructive atomic explosions. When a nuclear bomb detonates, it releases an enormous amount of energy in the form of heat and light, creating a blinding flash followed by a devastating shockwave. These immediate effects can cause death and destruction on a massive scale.

The Flash: The intense flash of a nuclear explosion is so bright that it can temporarily blind anyone who looks at it. Anyone within the blast zone can suffer severe eye damage.

Shockwave: The shockwave generated by a nuclear detonation can demolish buildings, vehicles, and structures, causing deaths and injuries directly.

High Temperature: The heat produced by a nuclear bomb

is sufficient to vaporize buildings and ignite anything flammable within the affected area.

Radiation and Fallout: Long-Term Effects

But the effects of nuclear weapons do not stop at the moment of the explosion. Nuclear weapons also release ionizing radiation, which can persist for a long time and have devastating effects on human health and the environment.

Ionizing Radiation: Ionizing radiation can damage living cells and cause genetic mutations that manifest in diseases like cancer and severe hereditary disorders in future generations.

Radioactive Fallout: After a nuclear explosion, radioactive particles can be carried by the air and settle on the ground, creating the so-called "radioactive fallout." This can contaminate the environment and food supply, causing further harm to human health.

Nuclear Winter: Global-Level Effects

One of the most terrifying aspects of nuclear weapons is their ability to trigger a "nuclear winter." This phenomenon is an indirect consequence of large-scale nuclear explosions, which release enormous quantities of particles into the atmosphere.

Global Darkening: Particles released into the atmosphere can block sunlight, causing significant global cooling. This darkening can influence the climate for months or even years, irreparably damaging crops and causing widespread famine.

Disturbed Environment: Nuclear winter can lead to drastic changes in the Earth's environment, including significant impacts on ecosystems, biodiversity, and the availability of natural resources.

The Fear of M.A.D. (Mutually Assured Destruction)

A central concept in nuclear warfare is the fear of mutually assured destruction (M.A.D.). This theory posits that if one nation were to launch a nuclear attack on another, the latter would respond with a nuclear counterattack, leading to complete mutual destruction. This fear helped maintain a precarious balance during the Cold War and deterred many direct conflicts between nuclear superpowers.

Conclusions

In summary, the catastrophic potential of nuclear weapons is an existential threat to life on Earth. Their immediate effects, long-term radiation, and the possibility of nuclear winter pose significant risks to human civilization and the environment. It is crucial that humanity continues to strive for nuclear disarmament and arms control to prevent the realization of such catastrophes. Only through the promotion of peace and diplomacy can we hope to avoid the disaster that nuclear weapons bring with them.

CHAPTER 3: FUKUSHIMA AND CHERNOBYL: LESSONS LEARNED

The nuclear accidents at Fukushima and Chernobyl represent two of the darkest chapters in the history of nuclear energy. These events demonstrated that, despite technological advances, the consequences of a nuclear catastrophe can be devastating and long-lasting. In this chapter, we will thoroughly examine the Fukushima and Chernobyl accidents, their causes, long-term consequences, and the crucial lessons that humanity has learned from them.

Chernobyl: The Mother of All Accidents

The Chernobyl nuclear accident, which occurred on April 26, 1986, is considered one of the worst nuclear disasters in human history. The Chernobyl nuclear power plant in Ukraine experienced a catastrophic explosion of reactor No. 4 during a safety test. This explosion released a massive amount of radioactive material into the atmosphere.

Causes: The Chernobyl accident was caused by a combination of human errors and design flaws. The crew did not follow safety protocols and failed to prevent the reactor from overheating, leading to the explosion.

Immediate Consequences: The explosion resulted in the deaths of two workers and exposed dozens of people to lethal doses of radiation. A large area around Chernobyl, known as the "exclusion zone," was evacuated and declared uninhabitable for decades.

Nuclear Winter and Long-Term Consequences: Although the Chernobyl accident did not cause a global nuclear winter, it had significant impacts on human health and the environment. Increased radiation-related illnesses, cancer, and genetic disorders were observed in exposed populations.

Fukushima: The Earthquake and Tsunami Incident

The Fukushima nuclear accident, which occurred on March 11, 2011, was triggered by a devastating earthquake and tsunami. The Fukushima Daiichi nuclear power plant in Japan was struck by an unusual wave that led to the failure of reactor cooling systems, resulting in overheating and core meltdowns.

Causes: The Fukushima accident was caused by an extreme natural event that exceeded the nuclear plant's resilience. The earthquake and tsunami damaged emergency systems, preventing reactor cooling.

Immediate Consequences: The evacuation of surrounding communities involved thousands of people, and radioactive materials were released into the environment. However, the number of direct radiation-related casualties was relatively low.

Long-Term Consequences: Fukushima highlighted challenges related to the long-term management of radioactive waste and the rehabilitation of affected areas. Concerns about nuclear safety led to a global review of existing nuclear facilities and a reevaluation of energy policies.

Lessons Learned

From Chernobyl and Fukushima, the following crucial lessons emerge:

Safety is Paramount: Nuclear safety must be a top priority. Accidents are often caused by human errors or design issues, and adequate training and oversight are essential for preventing catastrophes.

Emergency Management: Effective responses to nuclear emergencies are critical. Evacuation plans, emergency protocols, and clear communication strategies are necessary.

Transparency and Accountability: Authorities must be transparent in providing information to the public and be accountable for nuclear safety. Public trust is crucial.

Rethinking Nuclear Energy: The accidents at Fukushima and Chernobyl have led to a reevaluation of nuclear energy. Many nations have reduced or eliminated their reliance on this energy source.

Nuclear Waste Management: The long-term management of radioactive waste is a complex challenge that requires safe and sustainable solutions.

Conclusions

The nuclear accidents at Chernobyl and Fukushima remain tragic examples of the potentially catastrophic consequences of nuclear energy when safety is neglected, or unforeseen events surpass safety measures. These incidents have taught the world fundamental lessons about nuclear safety, transparency, and the need to carefully consider the risks associated with nuclear energy. For the future, humanity must commit to protecting the planet and future generations from such disasters by strengthening safety standards and seeking more sustainable energy solutions.

CHAPTER 4: THE HIDDEN COSTS OF NUCLEAR ENERGY

Nuclear energy is often seen as a clean and low-carbon energy source, but behind this positive image lie significant hidden costs. In this chapter, we will delve into the environmental and health costs associated with nuclear energy, highlighting impacts that may remain invisible but are profoundly relevant.

The Promise of Nuclear Energy

Nuclear energy has been considered a promising energy source as it does not directly produce greenhouse gas emissions during electricity generation. This characteristic has led many nations to invest in nuclear power plants as part of their strategies to address climate change.

The Nuclear Fuel Cycle and Radioactive Waste

However, the nuclear fuel cycle brings serious environmental and health issues. Nuclear energy production requires the conversion of uranium or plutonium into fissile material, a process that generates highly radioactive waste. This nuclear waste must be safely managed for thousands of years, posing a significant challenge in terms of safety and costs.

Nuclear Incidents and Their Fallout

Nuclear accidents, such as Chernobyl and Fukushima, highlight the human and environmental cost of nuclear catastrophes. Long-term consequences of these accidents include the contamination of agricultural lands and an increase in radiation-related diseases.

Managing Radioactive Waste

The issue of managing radioactive waste remains one of the most serious. The need to safely isolate these radioactive materials for thousands of years requires costly infrastructure and a long-term commitment from future generations.

Radioactive Emissions

Even operating nuclear power plants emit radioactive emissions into the environment. While they are strictly regulated, these emissions can still contribute to an increase in environmental radiation and potential human exposure.

The Risk of Nuclear Accidents

Although nuclear accidents are rare, their impact is so severe that the risk of one is considered unacceptable by many. The potential loss of human lives, environmental destruction, and the need for large-scale evacuations are all significant costs to consider.

The Role of the Nuclear Industry in Cost Management

The nuclear industry is responsible for managing many of these hidden costs. Research and development of waste management technologies, nuclear plant safety, and risk mitigation all require substantial investments.

The Alternative of Renewable Energies

For many, nuclear energy must be considered in light of alternatives. Renewable energies, such as solar and wind, are emerging as low-carbon energy sources without the same hidden costs as nuclear energy. Furthermore, renewables do not pose the risk of nuclear accidents or the thorny issue of radioactive waste.

Conclusions

Nuclear energy offers advantages in terms of low carbon emissions, but the associated hidden costs should not be underestimated. Managing radioactive waste, the risk of nuclear accidents, and radioactive emissions are all aspects that require careful consideration. In the face of increasingly competitive energy alternatives like renewables, it is important to carefully evaluate whether nuclear energy is the best choice for a sustainable future. Sustainability is not only about reducing carbon emissions but also about responsibly managing environmental and health risks.

CHAPTER 5: THE INFLUENCE OF THE NUCLEAR INDUSTRY ON POLITICS

The nuclear industry is intrinsically linked to politics in many nations. Economic interests, energy security, and environmental concerns often converge in the formulation of nuclear policies. In this chapter, we will explore the influence of the nuclear industry on politics, examining the connections between the industry, policymakers, and the decisions that shape the future of nuclear energy.

Lobbying and Political Donations

The nuclear industry is known for its lobbying efforts and political donations. Nuclear companies often seek to influence policymakers to gain support for new projects or protect existing interests. These efforts can influence policy formulation and legislative decisions.

The Nuclear Revolution: The 1950s and 1960s

During the 1950s and 1960s, the promotion of nuclear energy was supported by many governments as a solution to increasing energy demand and the need to reduce carbon emissions. Public investments and favorable policies stimulated the growth of the nuclear industry.

The Birth of Anti-Nuclearism

In the 1970s and 1980s, the nuclear industry faced growing public opposition. Nuclear accidents like Three Mile Island, Chernobyl, and increasing environmental awareness fueled anti-nuclear movements. These movements influenced public opinion and often compelled politicians to reconsider nuclear policies.

The Role of Regulatory Agencies

Regulatory agencies, such as the Nuclear Regulatory Commission (NRC) in the United States, play a crucial role in overseeing the nuclear industry. However, there is ongoing debate about their level of independence and their ability to ensure nuclear safety above industry interests.

Nuclear Disarmament and Foreign Policy

The nuclear industry not only concerns energy production but also foreign policy and global security. Nuclear-armed nations must balance the need for defense with the need to prevent a nuclear arms race. These considerations influence nuclear disarmament policies and international relations.

The Role of Universities and Research

Universities and research institutions often collaborate with the nuclear industry on research and the development of new technologies. These collaborations can influence the direction of scientific research and approaches to managing radioactive waste.

Political Decision-Making and Elections

Political elections are often a key moment for the nuclear industry. The positions of candidates and parties on nuclear issues can directly impact the future of nuclear policies and investments in the industry.

Lessons from Fukushima and Chernobyl

Nuclear accidents, like those at Fukushima and Chernobyl, have demonstrated how political decisions can have catastrophic consequences. These events have often pushed politicians to reevaluate their stance on nuclear energy and take stricter measures on nuclear safety.

The Challenge of Climate Change

Growing concerns about climate change have led some politicians to reconsider nuclear energy as a low-carbon energy source. The discussion of nuclear energy as part of the solution to climate change has reignited the debate on the role of the nuclear industry in energy policy.

Conclusions

The nuclear industry is inherently linked to politics in many nations. Its influence extends from the halls of power to research laboratories, from political elections to regulatory agencies. However, political decisions on nuclear energy are complex and often influenced by a range of factors, including economic interests, energy security, and environmental concerns. Nuclear policy must strike a balance between these aspects while considering the public good and safety.

CHAPTER 6: THE ROLE OF INTERNATIONAL ORGANIZATIONS IN NUCLEAR CONTROL

The peaceful use of nuclear energy and the prevention of nuclear weapon proliferation are crucial global objectives. To address these challenges, nations worldwide have established international organizations dedicated to nuclear control. In this chapter, we will examine the central role of the International Atomic Energy Agency (IAEA) and the Treaty on the Non-Proliferation of Nuclear Weapons (NPT) in ensuring global nuclear security and cooperation.

The International Atomic Energy Agency (IAEA)

The IAEA is an independent international organization based in Vienna, Austria, founded in 1957. Its primary mandate is to promote the peaceful use of nuclear energy and prevent the spread of nuclear weapons. The IAEA plays a fundamental role in various aspects of nuclear control:

- Inspections and Verification: The IAEA conducts regular inspections of nuclear facilities in NPT signatory countries to ensure they are not used for military purposes. This system of verifications contributes to deterring nuclear proliferation.
- Technical Assistance: The IAEA provides technical

assistance to developing countries to help them develop safe and peaceful nuclear programs. This assistance helps reduce technological disparities and promotes international cooperation.

- Radioactive Waste Management: The IAEA promotes best practices for the safe management of radioactive waste, contributing to preventing environmental contamination.
- Response to Nuclear Crises: The IAEA can play a crucial role in responding to nuclear crises, as it did during the Fukushima incident.

The Treaty on the Non-Proliferation of Nuclear Weapons (NPT)

The NPT is an international agreement aimed at preventing nuclear weapon proliferation and promoting nuclear disarmament. The treaty was signed in 1968 and was extended indefinitely in 1995. The NPT is based on three main pillars:

- Non-Proliferation: Signatory states commit not to seek the development or acquisition of nuclear weapons. In exchange, nuclear-armed states recognized by the treaty pledge to share nuclear technology for peaceful purposes.
- Nuclear Disarmament: Nuclear-armed signatory states commit to working toward complete nuclear disarmament, although significant progress in this direction has been slow.
- Peaceful Use of Nuclear Energy: Signatory states have the right to the peaceful use of nuclear energy, with the IAEA playing a crucial role in verifying peaceful use.

Criticism of the NPT and International Organizations

Despite the successes of the NPT and the IAEA, there are significant criticisms and challenges:

- Limited Nuclear Disarmament: Many argue that progress toward nuclear disarmament has been too slow, and NPT signatory nuclear states have not fulfilled their disarmament commitments.
- Proliferation by Non-Signatories: Countries like India, Pakistan, and Israel have developed nuclear weapons outside the NPT, raising doubts about its effectiveness.
- Access to Peaceful Use: Some countries have experienced restrictions on access to civilian nuclear energy, leading to tensions and controversies.
- Limited Resources: The IAEA is often constrained by financial and technical resources in its verification mandate.

The Importance of Dialogue and Cooperation

Despite the challenges, the NPT and the IAEA remain fundamental pillars of international nuclear control. Global security requires dialogue and cooperation among nations to address nuclear threats. International collaboration through organizations like the IAEA is essential to ensuring security and promoting a future without nuclear weapons.

Conclusions

The IAEA and the NPT are crucial organizations in international nuclear control. Preventing nuclear weapon proliferation and promoting the peaceful use of atomic energy require engagement and respect for international agreements. Despite challenges and criticisms, the importance of global dialogue and cooperation in the safe management of nuclear energy and the prevention of proliferation cannot be underestimated.

CHAPTER 7: THE ROLE OF INTERNATIONAL ORGANIZATIONS IN NUCLEAR CONTROL

In recent decades, nuclear disarmament has become an increasingly urgent goal for many global organizations and movements. In this chapter, we will examine the Global Nuclear Disarmament Initiative (GNDI) and analyze the movements and organizations working to reduce and ultimately eliminate nuclear weapons from the face of the Earth.

The Context of Nuclear Disarmament

Nuclear proliferation and the threat of nuclear weapons have led many people and organizations to seek ways to address this global danger. Nuclear disarmament has emerged as one of the most critical challenges of our time.

The Global Nuclear Disarmament Initiative (GNDI)

The Global Nuclear Disarmament Initiative is an international coalition of organizations and individuals working to promote global nuclear disarmament. The main goal of the GNDI is to mobilize public and political support for a binding global nuclear disarmament treaty.

The Treaty on the Prohibition of Nuclear Weapons

One of the milestones of the Global Nuclear Disarmament Initiative is the Treaty on the Prohibition of Nuclear Weapons

(TPNW), adopted by the United Nations General Assembly in 2017. The TPNW prohibits the development, production, possession, and use of nuclear weapons and has been ratified by a growing number of nations, although many nuclear-armed powers have not joined.

Movements for Nuclear Disarmament

The Global Nuclear Disarmament Initiative collaborates with a range of global movements for nuclear disarmament. These movements include non-governmental organizations, activists, religious associations, and many others.

The International Campaign to Abolish Nuclear Weapons (ICAN)

The International Campaign to Abolish Nuclear Weapons (ICAN) is a non-governmental organization that was awarded the Nobel Peace Prize in 2017 for its key role in promoting the TPNW. ICAN plays an essential role in raising awareness of the importance of nuclear disarmament and coordinating the activities of disarmament advocates worldwide.

The Anti-Nuclear Movement

The anti-nuclear movement has a long history dating back to the 1950s and 1960s when fear of nuclear weapons and protests against nuclear testing led to the creation of organizations like the Campaign for Nuclear Disarmament. These organizations have worked for years to raise public awareness and influence policies.

The Human Consequences of Nuclear Weapons

A fundamental part of the struggle for nuclear disarmament is highlighting the human consequences of nuclear weapons. Nuclear attacks would have a devastating impact on human life and the environment, causing death, illness, and widespread destruction.

Goals and Challenges of Nuclear Disarmament

The ultimate goal of nuclear disarmament is to create a world free of nuclear weapons. However, there are significant challenges in achieving this goal, including opposition from nuclear powers, national security concerns, and the technical complexity of reducing weapons.

The Role of Governments and Civil Society

Nuclear disarmament requires active involvement from both governments and civil society. Governments can join disarmament treaties and take concrete actions to reduce their nuclear arsenals, while civil society can play a key role in raising awareness and maintaining pressure on governments.

Conclusions

The Global Nuclear Disarmament Initiative and other movements for nuclear disarmament are making significant progress in raising awareness and mobilizing globally. Nuclear disarmament is a complex but crucial challenge for the future of global security. While the path to a world free of nuclear weapons is still long and challenging, the work done by these organizations and movements is essential for the well-being of humanity and the planet.

CHAPTER 8: EFFORTS FOR NUCLEAR WEAPONS REDUCTION

The issue of nuclear disarmament directly involves nuclear-armed powers, the nation's possessing nuclear arsenals. In this chapter, we will examine the crucial role these nations play in nuclear weapons reduction, exploring international agreements, reduction policies, and the challenges they face.

Unilateral Nuclear Weapons Reduction

Some nuclear-armed powers have taken unilateral actions to reduce their nuclear arsenals. These actions can include the destruction of obsolete weapons or the reduction in the number of nuclear warheads. These efforts can be a sign of commitment to nuclear disarmament but are often undertaken with the goal of modernizing their arsenals.

Nuclear Weapons Reduction Agreements

Many efforts for nuclear weapons reduction have been achieved through bilateral or multilateral agreements among nuclear-armed powers. Some of the most well-known agreements include:

- START I and START II: The Strategic Arms Reduction Treaty (START) agreements between the United States and the Soviet Union (later Russia) significantly reduced the number of strategic nuclear warheads and delivery vehicles.

- Strategic Arms Reduction Treaty III (START III): A proposed agreement aimed at further reducing the number of strategic nuclear warheads.
- Treaty on the Non-Proliferation of Nuclear Weapons (NPT): This treaty includes nuclear disarmament as one of its key pillars, although progress in this direction has been slow.
- Tactical Arms Reduction Treaty (TART): A proposed agreement to limit tactical nuclear weapons in Europe.

The Role of the United States and Russia

The United States and Russia possess the largest number of nuclear weapons globally and play a crucial role in reducing global nuclear arsenals. These two nations have made significant efforts to reduce the number of nuclear weapons, but there are significant challenges that remain, including geopolitical tensions and national security concerns.

Challenges to Nuclear Disarmament

Nuclear disarmament is complicated by several challenges:

- National Security: Nuclear-armed powers often argue that nuclear weapons are essential for their national security, making the disarmament process challenging.
- Budget and Priorities: Modernizing nuclear arsenals can be costly, and nations must balance military spending with other priorities.
- Geopolitical Tensions: Tensions among nuclear-armed powers can hinder disarmament efforts.

- Public Opinion: Public opinion can influence governments' willingness to engage in nuclear disarmament.

The Role of Civil Society

Civil society, including non-governmental organizations and activists, plays a fundamental role in supporting efforts for nuclear disarmament. These organizations raise public awareness, exert pressure on governments, and monitor the implementation of disarmament agreements.

Conclusions

Efforts for nuclear weapons reduction are a critical element for the future of global security. While many nations have taken actions to reduce their nuclear arsenals, significant challenges remain. Nuclear disarmament requires a combination of political commitments, public pressure, and international cooperation. The role of nuclear-armed powers, particularly the United States and Russia, is crucial for progress in this direction. Nuclear disarmament is a global challenge that requires the commitment of all those who desire a safer world free of nuclear weapons.

CHAPTER 9: THE CONTRIBUTION OF SCIENCE AND TECHNOLOGY TO NUCLEAR DISARMAMENT

Science and technology have played a crucial role in the development of nuclear weapons, but they can also play a fundamental role in promoting nuclear disarmament. In this chapter, we will examine how scientific research, technology, and innovation can contribute to the cause of nuclear disarmament, making the world safer and reducing the risk of nuclear conflicts.

- The Science of Deterrence

Nuclear deterrence, the theory that possessing nuclear weapons deters adversaries from attacking out of fear of destructive retaliation, has been a central part of nuclear strategy. Science and research have contributed to the development of theories and models that emphasize the dangers of nuclear deterrence, leading to a reconsideration of its effectiveness.

- Nuclear Verification

Science and technology play a crucial role in verifying disarmament agreements. Research has contributed to the development of more advanced monitoring and inspection technologies to ensure compliance with disarmament agreements. These tools include advanced sensors, imaging techniques, and radiation detection systems.

- Control of Nuclear Raw Materials

Science has contributed to the development of methods for monitoring nuclear raw materials such as uranium and plutonium to prevent them from falling into the wrong hands. These methods include isotopic signatures, which allow for tracking the origin of nuclear materials.

- Reduction of Nuclear Arsenals

Science and technology can play a key role in reducing nuclear arsenals. This can include the development of technologies for deactivating nuclear warheads or converting nuclear materials into less dangerous forms.

- Radioactive Waste Elimination

The management of radioactive waste is a critical part of nuclear disarmament. Scientific research can contribute to the development of safer and more efficient methods for managing and storing radioactive waste.

- The Humanitarian Impact of Nuclear Weapons

Scientific research has also contributed to highlighting the humanitarian impact of nuclear weapons, including the long-term consequences of nuclear explosions and the effects of radiation on human health. These findings have strengthened the case for nuclear disarmament.

- Artificial Intelligence and Nuclear Disarmament

Emerging technologies such as artificial intelligence (AI) can play a significant role in nuclear disarmament. AI can be used to analyze large amounts of data and identify violations of disarmament agreements more efficiently.

- The Commitment of the Scientific Community

The scientific community plays a crucial role in supporting nuclear disarmament. Many scientists and scientific organizations have signed petitions and declarations in favor of nuclear disarmament, lending their weight and credibility to the cause.

- The Ethical Responsibility of Science

Science and technology carry great ethical responsibility in the field of nuclear disarmament. Scientists have a duty to assess the impact of their discoveries on global security and promote responsible technology use.

- International Collaboration

Scientific research in the field of nuclear disarmament requires international cooperation. Scientists from different nations can collaborate to address common challenges and develop shared solutions.

Conclusions

Science and technology have played a fundamental role in the development of nuclear weapons, but they can also significantly contribute to the cause of nuclear disarmament. Scientific research can highlight the dangers of nuclear weapons, develop verification technologies, and contribute to the reduction of nuclear arsenals. However, progress in nuclear disarmament requires global commitment and the involvement of scientists, governments, and non-governmental organizations. Science can be a powerful force for peace, but political will is needed to translate scientific research into concrete actions for a world free of nuclear weapons.

CHAPTER 10: THE ROLE OF CHARISMATIC LEADERS IN NUCLEAR DISARMAMENT

Charismatic leaders have often played a crucial role in promoting nuclear disarmament. In this chapter, we will examine case studies of world leaders who have dedicated their commitment and influence to advance the cause of nuclear disarmament, demonstrating that leadership can be a crucial driver of change in this critical field.

• Mahatma Gandhi and India's Anti-Nuclearism

Mahatma Gandhi was an icon of non-violence and India's independence from British rule. His vision of a world without weapons, including nuclear arms, inspired many people worldwide. Gandhi advocated non-cooperation with nuclear energy and encouraged India to develop renewable energy sources, helping shape the country's nuclear policy.

• Mikhail Gorbachev and Glasnost

Soviet leader Mikhail Gorbachev played a crucial role in the nuclear disarmament process at the end of the Cold War. His policy of Glasnost contributed to a climate of greater openness and dialogue among nuclear superpowers. Gorbachev worked with

U.S. President Ronald Reagan to sign the Intermediate-Range Nuclear Forces Treaty (INF), eliminating an entire class of nuclear missiles.

- Nelson Mandela and Global Nuclear Disarmament

Nelson Mandela, an icon of the anti-apartheid struggle in South Africa, also promoted global nuclear disarmament. He supported the Nuclear Non-Proliferation Treaty (NPT) and called for the elimination of nuclear weapons. His moral commitment to peace inspired many people worldwide to support nuclear disarmament.

- Barack Obama and the Prague Speech

U.S. President Barack Obama delivered a historic speech in Prague in 2009, outlining a commitment to nuclear disarmament. He emphasized the importance of working toward a world without nuclear weapons and signed the New START Treaty with Russia, reducing limits on strategic nuclear warheads. While progress faced obstacles, Obama's speech underscored the importance of nuclear disarmament on the global political agenda.

- Setsuko Thurlow and the Voice of Hiroshima Survivors

Setsuko Thurlow, a survivor of the 1945 Hiroshima nuclear explosion, dedicated her life to sharing her experience and promoting nuclear disarmament. She worked with the International Campaign to Abolish Nuclear Weapons (ICAN) to raise public awareness and testified at numerous international conferences in support of the nuclear weapons ban.

- Pope Francis and the Hiroshima Declaration

Pope Francis visited Hiroshima in 2019 and issued a historic declaration condemning the use of nuclear weapons and emphasizing the importance of nuclear disarmament. He called on the world to "break the cycle of threat and fear" associated with nuclear arms and encouraged global dialogue for disarmament.

The Importance of Charismatic Leadership

Charismatic leadership can mobilize public opinion, inspire action, and influence political decisions. Charismatic leaders are often able to transform the nuclear disarmament debate from a technical issue into a moral and humanitarian cause. However, it is important to note that charismatic leadership alone is not sufficient; global commitment and concrete political support are needed to achieve nuclear disarmament.

Challenges and Opportunities

Promoting nuclear disarmament through charismatic leadership faces many challenges, including opposition from nuclear powers and the complexity of technical and political issues. However, it also offers a unique opportunity to mobilize public support and shape the political agenda. Charismatic leadership can inspire concrete actions, such as the signing of disarmament treaties or the involvement of governments and international organizations in the fight against nuclear weapons.

Conclusions

Charismatic leaders have played a fundamental role in promoting nuclear disarmament, contributing to raising public awareness, promoting disarmament agreements, and inspiring concrete actions. Their leadership demonstrates that nuclear disarmament is more than a political or technical issue; it is a moral and humanitarian cause that involves the future of global security and human survival. History teaches us that charismatic leadership can make a difference in the struggle for a world without nuclear weapons.

CHAPTER 11: PEACE AND NUCLEAR DISARMAMENT MOVEMENTS WORLDWIDE

Across the globe, a myriad of movements and organizations are dedicated to peace and nuclear disarmament. In this chapter, we will examine some of these movements and the challenges they face as they strive to promote a world free of nuclear weapons.

- International Campaign to Abolish Nuclear Weapons (ICAN)

ICAN is a non-governmental organization that was awarded the Nobel Peace Prize in 2017 for its crucial role in promoting the Treaty on the Prohibition of Nuclear Weapons (TPNW). ICAN coordinates the efforts of partner organizations worldwide to raise public awareness and advocate for nuclear disarmament.

- Mayors for Peace

Mayors for Peace is a global network of mayors and local governments working to promote peace and nuclear disarmament. Founded in 1982 by Hiroshima and Nagasaki, the organization now has thousands of member cities committed to promoting nuclear disarmament at the local level.

- Global Zero

Global Zero is an international organization dedicated to the complete elimination of nuclear weapons. It works to engage world leaders, experts, and activists in promoting nuclear disarmament.

- International Physicians for the Prevention of Nuclear War (IPPNW)

IPPNW is an international medical organization that was awarded the Nobel Peace Prize in 1985. It advocates for the elimination of nuclear weapons, highlighting the catastrophic consequences of nuclear explosions on human health.

- Pugwash Conferences on Science and World Affairs

The Pugwash Conferences are international meetings of scientists, academics, and political leaders focusing on issues of peace and nuclear disarmament. These conferences promote dialogue and international cooperation to address nuclear challenges.

- Women's International League for Peace and Freedom (WILPF)

WILPF is one of the oldest organizations dedicated to peace promotion. Women have played a crucial role in the nuclear disarmament movement, bringing a justice-based and conflict-prevention approach.

- Abolition 2000

Abolition 2000 is a global network of civil society organizations working for nuclear disarmament. The organization seeks to mobilize public support and influence governments globally.

- Campaign for Nuclear Disarmament (CND)

CND is a British organization that has advocated for nuclear disarmament since 1958. It has organized protests and demonstrations, influencing the nuclear policy of the United Kingdom.

- The Bulletin of the Atomic Scientists

The Bulletin is a publication providing analysis and information on the nuclear threat and global security issues. The Bulletin's "Doomsday Clock" symbolically represents humanity's proximity to a nuclear catastrophe.

- Challenges Faced by Peace and Nuclear Disarmament Movements

Peace and nuclear disarmament movements face numerous challenges, including opposition from nuclear powers, the complexity of technical and political issues, and public apathy. However, these movements play a crucial role in drawing global attention to the issue of nuclear disarmament.

Conclusions

Peace and nuclear disarmament movements are a fundamental force for change. They work to raise public awareness, influence policies, and promote global dialogue on nuclear weapons. While the challenges are significant, the importance of their mission cannot be underestimated. They are a critical voice in the struggle for a safer world free of nuclear weapons and represent hope for a future where the nuclear threat is only a memory of the past.

CHAPTER 12: THE PERSPECTIVE OF NUCLEAR VICTIMS

No analysis of nuclear issues and their impact can disregard the voices of direct victims and eyewitnesses of nuclear incidents. In this chapter, we will listen to the poignant stories of survivors of nuclear incidents and those who have personally witnessed the devastating consequences of nuclear events.

Hiroshima and Nagasaki: Voices from the Past

The atomic bombings of Hiroshima and Nagasaki in 1945 resulted in the immediate death of tens of thousands and inflicted unimaginable suffering on survivors. The testimonies of Hiroshima and Nagasaki survivors bear vivid witness to the destructive power of nuclear weapons.

Katsuji Yoshida, Hiroshima Survivor: Yoshida was 13 years old when Hiroshima was hit by the atomic bomb. He recounted being engulfed by flames and debris as he tried to help victims. His story serves as a warning against nuclear weapons.

Setsuko Thurlow, Hiroshima Survivor: Setsuko Thurlow was a 13-year-old student when Hiroshima was bombed. She lost family and friends in the blast and has testified worldwide to raise public awareness about the human consequences of nuclear weapons.

Chernobyl: Witnesses to Nuclear Hell

The explosion of the Chernobyl nuclear reactor in 1986 led

to one of the worst nuclear disasters in history. Survivors and rescue workers faced high radiation levels and extreme conditions.

Lyudmila Ignatenko, Wife of a Chernobyl Firefighter: Lyudmila's husband, Vasily Ignatenko, was one of the firefighters who responded to the Chernobyl fire. He suffered severe radiation sickness and endured terrible suffering before his death. Lyudmila's testimony tells of the desperation and courage of those confronting the aftermath of the disaster.

Valery Legasov, Chernobyl Scientist: Valery Legasov was one of the leading scientists involved in managing the Chernobyl incident. He worked tirelessly to address the crisis and later testified about the lessons learned from the incident. His death, an apparent suicide, became a symbol of the challenges faced by those who fought against the nuclear disaster.

Fukushima: The Resilience of the Japanese People

The Fukushima nuclear incident in 2011 was triggered by an earthquake and tsunami, leading to a series of nuclear reactor meltdowns. Those affected by this disaster displayed extraordinary resilience.

Naoto Matsumura, The Last Inhabitant of Fukushima: After the evacuation of Fukushima, Naoto Matsumura remained in the contaminated area to care for abandoned animals. His story reflects his love for his homeland and determination to confront the challenges of nuclear contamination.

The Mothers of Fukushima: A group of Japanese mothers joined together to demand transparency regarding the health consequences of their children following the Fukushima incident. Their protests brought attention and support to the cause of nuclear disarmament.

Conclusions

The stories of nuclear incident survivors and eyewitnesses remind us of the humanity behind statistics and nuclear policies. These individuals have endured the worst consequences of nuclear events but have also demonstrated extraordinary strength and determination in raising public awareness and promoting nuclear disarmament.

Listening to these voices urges us to reflect on the human cost of nuclear weapons and the urgency of working toward a world without them. Their testimonies invite us to consider nuclear issues not only as political or technical matters but as humanitarian challenges that require a global response. It is the voices of nuclear victims that should guide us in our commitment to a safer, nuclear-free future.

CHAPTER 13:
THE ROLE OF ART AND CULTURE IN PROMOTING NUCLEAR DISARMAMENT

Art and culture are powerful tools for conveying messages, inspiring empathy, and raising public awareness. In this chapter, we will examine how art and culture can contribute to promoting nuclear disarmament by providing a creative and engaging perspective on this critical global issue.

- Art as a Reflection of Nuclear Consequences

Artists from around the world have used various forms of art to depict the consequences of nuclear weapons. Paintings, photographs, sculptures, and literary works have captured the destruction, pain, and suffering caused by nuclear incidents such as Hiroshima, Nagasaki, Chernobyl, and Fukushima. These works provide an emotional testimony to the horrors of nuclear weapons and can evoke empathy and compassion in viewers.

- Music as a Vehicle for Peace Messages

Music has the power to touch people's emotions and convey messages of peace and nuclear disarmament. Songs like John Lennon's "Imagine" or Nena's "99 Red Balloons" have addressed nuclear themes and the fear of nuclear war. Peace concerts and music festivals dedicated to nuclear disarmament have been organized worldwide to unite people in the fight against nuclear weapons.

- Cinema as a Narrative Medium

Cinema has produced a wide range of films that explore nuclear themes, from realistic portrayals of the effects of nuclear weapons to fictional tales of apocalyptic scenarios. Films like "The Day After" and "Threads" have depicted the devastating consequences of nuclear warfare and helped raise concerns about nuclear safety.

- Theater as an Awareness Tool

Theater has been used to create plays and performances that address nuclear issues and nuclear disarmament. These performances can engage the audience intimately, prompting them to reflect on the implications of nuclear weapons. Political and social theater has tackled nuclear issues as a matter of urgent importance.

- Literature as the Voice of Conscience

Novelists and poets have explored nuclear themes in their literary works, often highlighting the human aspects of nuclear-related stories. Works like Ryunosuke Akutagawa's "Sabbie mobili" and John Irving's "The World According to Garp" have approached nuclear themes in different ways, contributing to stimulating reflection and discussion.

- Art as a Tool of Protest

Around the world, committed artists have used art as a means of protest against nuclear weapons. Murals, artistic installations, and public performances have been employed to draw attention to nuclear disarmament and promote collective action. These forms of creative expression can have a lasting impact on public awareness.

- The Role of Artists and Intellectuals

Artists and intellectuals have often played a crucial role

in promoting nuclear disarmament. Their influential voices can raise public awareness and compel political leaders to take concrete actions. Art and culture can provide a unique perspective and inspire social change.

- Challenges of Art and Culture in Promoting Nuclear Disarmament

Despite the potential of art and culture in promoting nuclear disarmament, there are challenges to overcome. Art can be subjective and open to various interpretations, and its effectiveness in conveying specific messages may vary. Furthermore, reaching a broad and diverse audience through art and culture can be challenging.

Conclusions

Art and culture play a fundamental role in promoting nuclear disarmament, conveying emotional and engaging messages that touch the hearts and minds of people. These forms of art can inspire empathy, raise public awareness, and push for action. Art and culture are valuable means of raising awareness and promoting reflection on the nuclear threat, inviting the world to seriously consider the path toward a nuclear-free future.

CHAPTER 14: THE VISION OF A WORLD WITHOUT NUCLEAR WEAPONS

Let's imagine a future where nuclear weapons are no longer part of the equation in global security. In this chapter, we will explore the vision of a world without nuclear weapons, the advantages of a peaceful world, and the challenges to achieving this dream.

The Promise of a World Without Nuclear Weapons

A world without nuclear weapons is a shared vision among many leaders, organizations, and individuals worldwide. This vision is based on the belief that peace and security can be achieved through non-nuclear means. Envisioning a world without nuclear weapons involves:

- Security Based on Other Means

A world without nuclear weapons requires strengthening other forms of security, such as diplomacy, dialogue, international cooperation, and conflict prevention. States must learn to resolve disputes through negotiation and respect for international norms.

- Reduction of International Tensions

Without nuclear weapons, tensions between nations can significantly decrease. The fear of sudden nuclear warfare disappears, opening the door to greater collaboration on global issues such as climate change, poverty, and health.

- Respect for Human Life

A world without nuclear weapons is one where human life is considered sacred and inviolable. Nuclear weapons carry the threat of indiscriminate destruction, but without them, humanity can live without the constant fear of a nuclear catastrophe.

Advantages of a World Without Nuclear Weapons

Imagining a world without nuclear weapons is not only an ideal vision but also a practical perspective that offers many advantages:

- Reduction of the Risks of Accidental Nuclear War

The removal of nuclear weapons eliminates the possibility of accidental nuclear war caused by calculation errors or technical malfunctions. This drastically reduces the likelihood of nuclear conflict.

- Release of Economic Resources

The expenditure on maintaining, upgrading, and modernizing nuclear weapons is a significant economic burden for states. A world without nuclear weapons frees up financial resources to address more urgent challenges, such as poverty and climate change.

- Prevention of Nuclear Proliferation

A world without nuclear weapons makes nuclear proliferation less attractive. The reduced nuclear threat can deter other countries from seeking to develop nuclear weapons, contributing to global stability.

- Possibility of Sustainable Human Development

Peace is a prerequisite for sustainable human development. Without the constant threat of nuclear weapons, global efforts can be focused on urgent issues like access to water, education, health, and justice.

Challenges in Achieving the Dream

However, achieving a world without nuclear weapons is not without challenges:

- Opposition from Nuclear Powers

Current nuclear powers have a strategic interest in retaining their nuclear weapons and may actively oppose efforts for nuclear disarmament. Diplomacy and dialogue are crucial for engaging them in the cause of disarmament.

- International Security

Resolving regional and international tensions is essential for creating a world without nuclear weapons. States must work together to address security issues and resolve conflicts peacefully.

- Monitoring and Control

Ensuring that states adhere to nuclear disarmament agreements requires rigorous international monitoring and verification. Organizations like the International Atomic Energy Agency (IAEA) play a crucial role in this process.

- Public Awareness

Educating the public about the importance of nuclear disarmament is crucial. Art, culture, peace movements, and the voices of nuclear victims can play a critical role in engaging the public and driving change.

The Reality of a World Without Nuclear Weapons

Although achieving a world without nuclear weapons is a challenge, it is a challenge worth facing. Humanity has demonstrated its ability to overcome significant obstacles throughout history. A world without nuclear weapons is one where peace, cooperation, and respect for human life are at the forefront of the global agenda.

Let us always remember that the vision of a world without nuclear weapons is a shared goal of many people worldwide. Working together to realize it is not only a responsibility but also a promise of a safer and more peaceful future for future generations.

CHAPTER 15: GEOPOLITICAL TENSIONS AND NUCLEAR DISARMAMENT

Nuclear disarmament is a complex challenge, and geopolitical tensions play a significant role in its progress. In this chapter, we will examine the main geopolitical tensions that hinder nuclear disarmament and how these challenges can be addressed to promote a vision of a world without nuclear weapons.

- The Role of Nuclear Powers

The most obvious geopolitical tensions in the context of nuclear disarmament concern the nuclear powers themselves. States with nuclear arsenals, such as the United States, Russia, China, the United Kingdom, and France, have a strategic interest in maintaining their nuclear weapons. They see these weapons as a deterrent against threats to their security and are unwilling to unilaterally give them up.

- The Theory of Nuclear Weapons-Based Security

One of the primary tensions is the belief that nuclear weapons ensure a state's security. This theory, known as "Nuclear Weapons-Based Security Theory," argues that the threat of nuclear weapon use deters adversaries and prevents large-scale conflicts. Many states believe that nuclear disarmament could weaken their security position, making it difficult for them to engage in the disarmament process.

- Nuclear Proliferation

Another source of geopolitical tension is nuclear proliferation. The concern is that if current nuclear powers reduce or relinquish their nuclear weapons, other countries may be incentivized to develop their nuclear capabilities. This could increase the number of nuclear actors, raising the risk of accidents or unauthorized use of nuclear weapons.

- Regional Tensions

Regional tensions between states, often fueled by territorial disputes or historical rivalries, represent another significant obstacle to nuclear disarmament. These tensions make it challenging to convince states to give up their nuclear weapons if they perceive a real or potential threat in their region.

- Lack of a Global Security Framework

The absence of an effective global security framework is a constant tension in the nuclear disarmament process. The lack of binding multilateral agreements, the inefficiency of existing international organizations, and the absence of mechanisms for resolving complex global conflicts hinder disarmament efforts.

- Lack of Trust Among Parties

Geopolitical tensions often result in a lack of mutual trust among the parties involved in nuclear disarmament. The fear that other states may not abide by disarmament agreements or may seek strategic advantages through deception undermines international cooperation.

- National Interests and Domestic Politics

National interests and domestic politics play a significant role in the decisions of political leaders regarding nuclear disarmament. Decisions on disarmament can be influenced by domestic political objectives, including national security, political popularity, and legitimacy.

Addressing Geopolitical Tensions for Nuclear Disarmament

While geopolitical tensions pose a significant challenge to nuclear disarmament, there are approaches that can be adopted to address them:

- Active Diplomacy

Active diplomacy and open dialogue between nuclear powers and countries interested in nuclear disarmament are essential. States must seek to understand each other's concerns and work to find shared solutions.

- Multilateral Agreements

Binding multilateral agreements can help establish common norms and commitments for nuclear disarmament. For example, the Treaty on the Non-Proliferation of Nuclear Weapons (NPT) provides a framework for nuclear arms control and disarmament.

- Building Trust

Building trust among the involved parties is crucial. This can be achieved through transparent implementation of agreements, cooperation on shared security issues, and adherence to commitments.

- Involvement of Civil Society and International Organizations

Civil society and international organizations can play a significant role in promoting nuclear disarmament. They can raise public awareness, promote transparency, and monitor compliance with agreements.

- Providing Incentives

Incentivizing states to give up nuclear weapons through alternative security measures, security assurances, and economic benefits can help overcome resistance to disarmament.

Conclusion

Geopolitical tensions represent a significant challenge for nuclear disarmament, but they are not insurmountable. Through diplomacy, trust-building, multilateral agreements, and the involvement of civil society, it is possible to address these tensions and make progress toward a world without nuclear weapons. Global peace and security depend on nations' ability to work together to address this critical challenge.

CHAPTER 16: THE IMPORTANCE OF NUCLEAR ISSUE EDUCATION

Education plays a fundamental role in promoting awareness of the nuclear issue and shaping future perspectives on nuclear matters. In this chapter, we will explore the importance of nuclear issue education and how it can contribute to progress toward a world without nuclear weapons.

- Understanding the Complexity of Nuclear Matters

Nuclear issues are among the most complex and technical challenges of our time. To address this complexity, education plays a key role in providing accurate and understandable information about nuclear weapons, nuclear energy, and related challenges. Students, citizens, and policymakers must grasp the fundamental concepts of nuclear issues to participate knowledgeably in the disarmament debate.

- Promoting Awareness of Consequences

Education on the nuclear issue can vividly illustrate the human, environmental, and geopolitical consequences of nuclear weapons and nuclear incidents. Knowledge of the devastating consequences of Hiroshima, Nagasaki, Chernobyl, and Fukushima can evoke empathy and mobilize public opinion against nuclear weapons.

- Developing Critical Thinking Skills

Education on the nuclear issue goes beyond providing data and information; it also aims to develop critical thinking skills. Students should be able to critically evaluate nuclear policies, examine arguments for and against nuclear weapons, and engage in informed discussions.

- Promoting Nuclear Disarmament

A key objective of nuclear issue education is to promote nuclear disarmament. Educators can emphasize the importance of concrete commitments to reduce nuclear arsenals, ratify disarmament treaties, and work toward a world free of nuclear weapons. Education can inspire the next generation of peace activists and advocates for nuclear disarmament.

- Involving Civil Society

Education on the nuclear issue involves not only students in classrooms but also civil society as a whole. Non-governmental organizations, academic institutions, and citizen groups can organize conferences, seminars, workshops, and awareness campaigns on the nuclear issue. Education thus becomes a means of social mobilization.

- Valuing Scientific Skills

Education on the nuclear issue can emphasize the scientific and technological skills needed to address nuclear matters responsibly. Students can learn the principles of nuclear physics, radiological protection, and nuclear technologies, contributing to the training of experts in the field.

- Focusing on the Future

Education on the nuclear issue should focus on building a safer future. Educators can challenge students to imagine a world without nuclear weapons and consider the actions necessary to achieve it. This future perspective can inspire commitment to nuclear disarmament.

- Providing Accessible Resources

It is essential that educational resources on the nuclear issue are accessible to all. This includes free educational materials, online courses, documentaries, and multilingual resources. Democratizing nuclear issue education allows a broader audience to engage in the debate.

- Engaging Future Leaders

Educators can play a crucial role in encouraging young people to consider careers addressing nuclear challenges, such as nuclear diplomacy, international security, and scientific research. Preparing future leaders to address nuclear issues is vital for progress toward disarmament.

- International Collaboration

Education on the nuclear issue should be a global effort, involving educators, academic institutions, and non-governmental organizations from around the world. International collaboration can help spread best practices and share resources.

Conclusion

Education on the nuclear issue is a fundamental pillar in promoting nuclear disarmament. It provides the knowledge, skills, and awareness necessary to address nuclear matters responsibly. Educators, civil society, and individuals have a crucial role in shaping the future of nuclear issues and working toward a safer world free of nuclear weapons.

CHAPTER 17: THE ROLE OF NON-GOVERNMENTAL ORGANIZATIONS IN NUCLEAR DISARMAMENT

Non-Governmental Organizations (NGOs) play a crucial role in advocating for nuclear disarmament. In this chapter, we will examine the impact of NGOs in promoting nuclear disarmament, the strategies they employ, and the results they have achieved.

- NGO Commitment to Nuclear Disarmament

NGOs engaged in nuclear disarmament focus on a wide range of activities to raise public awareness, influence policymakers, and promote change. These activities include public awareness campaigns, advocacy, research, education, and policy monitoring.

- Public Awareness Campaigns

NGOs launch public awareness campaigns to inform the public about the threats posed by nuclear weapons and to garner support for disarmament. They use mediums such as social media, online petitions, public events, and art installations to engage the public and create awareness.

- Advocacy with Policymakers

NGOs work to influence policymakers at both national and international levels. They engage in lobbying, meet with legislators, and participate in international conferences to promote disarmament policies and monitor compliance with existing agreements.

- Research and Analysis

Research-oriented NGOs produce detailed analyses of nuclear issues, including the costs and risks of nuclear weapons, the effectiveness of disarmament policies, and the human and environmental implications of nuclear incidents. Informed research provides solid data to support the disarmament cause.

- Education and Training

Many NGOs provide educational and training programs for students, activists, and emerging leaders. These programs help develop critical skills and awareness of nuclear issues among new generations.

- Monitoring and Verification

Some NGOs engaged in nuclear disarmament play a crucial role in monitoring compliance with international agreements and the nuclear policies of states. This contributes to transparency and accountability in the nuclear field.

- International Collaboration

NGOs often work together in global networks and coalitions to amplify their influence. Notable examples include the International Campaign to Abolish Nuclear Weapons (ICAN) and the 2017 Nobel Peace Prize awarded to ICAN for its work in promoting the Treaty on the Prohibition of Nuclear Weapons.

- Achievements of NGOs in Nuclear Disarmament

NGOs have achieved significant successes in the field of nuclear disarmament. One of the most notable accomplishments is the adoption of the Treaty on the Prohibition of Nuclear Weapons in 2017, thanks to the efforts of ICAN and other NGOs. This treaty

represents a significant step toward a world without nuclear weapons, although it has not been adopted by all nuclear-armed powers.

- Challenges Faced by NGOs

NGOs engaged in nuclear disarmament face numerous challenges, including opposition from nuclear-armed powers and their allies, lack of financial resources, and policies hostile to their goals. However, these challenges have not deterred their commitment.

- The Crucial Role of NGOs in Advancing Toward a Nuclear-Free World NGOs play a crucial role in promoting nuclear disarmament and keeping attention

focused on nuclear threats. Their work contributes to pushing governments to take concrete steps toward a world without nuclear weapons.

Conclusion

Non-Governmental Organizations are agents of change in the field of nuclear disarmament. Through awareness campaigns, advocacy, research, and monitoring, NGOs promote awareness and drive political change. Their commitment is essential for progress toward a safer world free of nuclear weapons, and their work continues to inspire activists, students, and citizens worldwide to support this critical cause.

CHAPTER 18: INDIVIDUAL RESPONSIBILITY FOR NUCLEAR DISARMAMENT

Nuclear disarmament is not just a matter of global politics but also a responsibility that every individual can take on. In this chapter, we will explore the role of individual responsibility in supporting nuclear disarmament and the concrete actions that people can undertake to promote this vital cause.

- The Importance of Awareness

Awareness is the first step towards supporting nuclear disarmament. Individuals must educate themselves about nuclear threats, the consequences of nuclear arsenals, and the challenges related to nuclear issues. Research and learning are fundamental to fully understand the issue.

- Engaging the Public

Individuals can play a crucial role in engaging the public in discussions about nuclear disarmament. This can include participating in protests, sharing information on social networks, writing opinion pieces, and taking part in awareness-raising events.

- Communicating with Political Leaders

Individuals can write to their political representatives, urging them to commit to nuclear disarmament. Citizen influence on policy is a key element of democracy and can help push governments towards concrete actions.

- Supporting Disarmament Organizations and Movements

People can financially and voluntarily support organizations and movements dedicated to nuclear disarmament. These organizations work to raise public awareness, influence policies, and promote concrete actions.

- Promoting the Treaty on the Prohibition of Nuclear Weapons

Individuals can support and promote the Treaty on the Prohibition of Nuclear Weapons (TPNW) adopted in 2017. Even though some nuclear-armed powers have not ratified it, the TPNW represents a significant step towards disarmament and has the support of many countries.

- Education and Training

Individuals can participate in courses, seminars, and workshops on nuclear issues. These opportunities offer a deeper understanding of the matter and can provide skills for effective support of nuclear disarmament.

- Promoting Dialogue and Diplomacy

Individuals can promote dialogue and diplomacy as means of resolving international disputes. Using dialogue instead of nuclear threats can help reduce global tensions.

- Reducing Dependence on Nuclear Energy Sources

Reducing dependence on nuclear energy sources can indirectly contribute to disarmament. Individuals can support the use of renewable and sustainable energies as alternatives to nuclear energy.

- Engaging New Generations

Involving young people in discussions on nuclear disarmament is crucial for the future. Individuals can work with schools and institutions to organize educational programs and awareness campaigns among youth.

- Responsibility in Social Resource Use

People can exercise their individual responsibility in the use of social resources. This means sharing accurate and responsible information on nuclear issues, avoiding misinformation, and promoting constructive dialogue.

- Engaging in Elections

Participating in elections and supporting candidates who promote nuclear disarmament is a direct way to influence nuclear policies. Individuals can vote for leaders who prioritize nuclear disarmament on their agenda.

- Being Role Models

Individuals can be role models for others through their active commitment to nuclear disarmament. This can inspire others to take a stand and follow their example.

Conclusion

Individual responsibility in supporting nuclear disarmament is an essential component for progress toward a world free of nuclear weapons. Each individual can contribute to the cause in different ways, from public awareness to advocacy, from participating in awareness events to education and training. Nuclear disarmament is a global challenge, but it requires the commitment of each of us to achieve a safer and more peaceful future for generations to come.

CHAPTER 19: THE ROADMAP TO A NUCLEAR-FREE WORLD

Achieving a world without nuclear weapons is an ambitious but attainable goal. To reach this crucial outcome, it is necessary to define a roadmap with concrete steps and well-planned measures. In this chapter, we will explore a roadmap for nuclear disarmament, highlighting the actions required to achieve this objective.

- Step 1: Global Commitment

The first fundamental step is to secure global commitment to nuclear disarmament. This requires the participation of all nations, including nuclear-armed countries. A significant starting point is the universal ratification of the Treaty on the Prohibition of Nuclear Weapons (TPNW), which provides a legal basis for disarmament.

- Step 2: Reduction of Nuclear Arsenals

Nuclear powers must commit to the gradual reduction of their nuclear arsenals. This process should be conducted transparently and verifiably, with the goal of reducing nuclear stockpiles to credible minimum levels.

- Step 3: Verification and Transparency

Verification and transparency are crucial for nuclear

disarmament. Nations must allow international inspection and verification of their nuclear facilities to ensure compliance with disarmament commitments.

- Step 4: Elimination of Tactical Weapons

Tactical nuclear weapons, often overlooked but highly dangerous, must be specifically addressed. States should commit to the gradual elimination of these weapons, thereby reducing the risk of accidental use.

- Step 5: Conversion of Nuclear Infrastructure

Nuclear infrastructures should be converted for peaceful purposes. Uranium enrichment facilities and plutonium production facilities can be repurposed for civilian nuclear uses, such as energy production or medicine.

- Step 6: Security and Custody

The security and custody of existing nuclear weapons must be a priority. This means ensuring that these weapons are protected from theft, acts of terrorism, or unauthorized access.

- Step 7: Involvement of International Organizations

International organizations, especially the International Atomic Energy Agency (IAEA), must play a central role in overseeing nuclear disarmament. The IAEA should receive a broader mandate to monitor and verify disarmament.

- Step 8: Education and Awareness

Education and awareness play a crucial role in the roadmap. Efforts to engage civil society, educate young people, and raise public awareness about nuclear issues are essential to support nuclear disarmament.

- Step 9: Multilateral Diplomacy

Multilateral diplomacy is essential for nuclear disarmament. International conferences, negotiations, and agreements should be used to promote disarmament and resolve nuclear disputes.

- Step 10: Alternative Security Measures

Nations should develop alternative security measures to ensure their defense without relying on nuclear weapons. This may include security agreements, regional military cooperation, and strengthening conventional defense capabilities.

- Step 11: Universal Ratification of the TPNW

Universal ratification of the Treaty on the Prohibition of Nuclear Weapons should be a final goal. This treaty commits states not to develop, test, produce, acquire, possess, or stockpile nuclear weapons.

- Step 12: Ongoing Monitoring and Review

The roadmap for nuclear disarmament should be subject to ongoing monitoring and periodic review. Emerging challenges and changes in the international context require constant adaptation of strategies.

- Step 13: Strengthening International Organizations

International organizations need to be strengthened to play a more effective role in nuclear disarmament. This may include a review of governance structures and increased resources.

- Step 14: Development of Global Peace Structures

Nuclear disarmament should be seen within a broader framework of global peace. Nations must work together to address the underlying causes of conflicts and promote international cooperation.

- Step 15: Engagement of New Generations

Engaging new generations is crucial for the long-term success of nuclear disarmament. Efforts to educate and raise awareness among young people must be supported and reinforced.

Conclusion

The roadmap to a world without nuclear weapons requires constant commitment, international cooperation, and concrete actions. Achieving this goal will not be easy, but it is a moral and strategic imperative. Nuclear disarmament can contribute to creating a safer and more peaceful world for all inhabitants of the planet, and the roadmap outlined in this chapter provides guidance for pursuing this important objective.

CHAPTER 20: DREAM OR REALITY? THE FUTURE OF NUCLEAR DISARMAMENT

In the final chapter of this discourse on nuclear disarmament, we will explore the future of the nuclear disarmament movement and reflect on the challenges and opportunities that await those who pursue the goal of a world free of nuclear weapons.

- The Vision of a Nuclear-Free World

The vision of a world without nuclear weapons is a source of inspiration for many but remains an ambitious goal. However, it is important to remember that even the greatest challenges can be overcome when there is global commitment.

- Challenges Persist

Despite progress in the nuclear disarmament movement, challenges persist. Some nuclear-armed powers have increased their spending on arsenal modernization, while others have resisted joining the Treaty on the Prohibition of Nuclear Weapons (TPNW). Geopolitical tensions and security concerns continue to hinder disarmament.

- The Need for Global Leadership

Global leadership is essential for the success of nuclear disarmament. Nuclear-armed nations must take responsibility for leading the disarmament process and demonstrating that security can be ensured without nuclear weapons.

- Involvement of Civil Society

Civil society plays a fundamental role in promoting nuclear disarmament. Activists, NGOs, and citizens must continue to call for government action and raise public awareness about the consequences of nuclear weapons.

- The Power of Diplomacy

Diplomacy remains a powerful tool for nuclear disarmament. Diplomatic meetings, negotiations, and international agreements can help reduce tensions and promote trust among nations.

- Technology and Security

Technology plays a crucial role in nuclear security. Innovations in conventional weapons, cyber threats, and missile defense systems can influence the context of nuclear disarmament.

- Verification and Transparency

Verification and transparency remain challenges in nuclear disarmament. Efforts to ensure access to and verification of nuclear facilities must be intensified.

- The Role of International Organizations

International organizations, such as the IAEA, must continue to play a central role in nuclear disarmament. These organizations can help ensure compliance with nuclear agreements.

- The Role of Nuclear Powers

Nuclear powers must consider nuclear disarmament a strategic priority. Reducing nuclear arsenals and engaging in multilateral negotiations are fundamental steps.

- Sustainability of the Nuclear Disarmament Movement

The nuclear disarmament movement must remain sustainable in the long term. This requires active involvement from new generations and maintaining focus on the goal of disarmament.

- The Influence of Public Opinion

Public opinion can exert significant influence on nuclear policies. Individuals who mobilize and voice their opinions can push governments to take concrete steps toward disarmament.

- Opportunities in Nuclear Disarmament

Despite the challenges, there are also opportunities in the field of nuclear disarmament. Change can occur when a confluence of favorable factors exists, and political will, diplomacy, and public pressure can create conditions for progress.

- The Urgency of Nuclear Disarmament

The urgent need for nuclear disarmament cannot be emphasized enough. Nuclear weapons pose an existential threat to humanity and the planet itself. Disarmament is not an option but a necessity.

- Responsibility for All

The responsibility for nuclear disarmament belongs to all of us. Every individual, community, and nation has a role to play in pursuing this critical cause.

Conclusion

The future of nuclear disarmament is uncertain, but the vision of a world without nuclear weapons must continue to guide our efforts. Disarmament is a global challenge that requires the collaboration of everyone, from political leadership to civil society and the new generations. We cannot afford to postpone action. The future of our planet and civilization depends on our commitment to a world free of nuclear weapons. The dream of a nuclear-free world can become a reality if we all commit to making it happen.

CHAPTER 21: PERSONAL OPINIONS AND INNOVATIVE IDEAS FOR NUCLEAR DISARMAMENT

In addition to examining the challenges and opportunities for nuclear disarmament, it is important to share some personal opinions and innovative ideas for addressing the issue more effectively.

- Promoting Preventive Diplomacy

An innovative approach could be to promote preventive diplomacy as a means to prevent conflicts that could lead to the use of nuclear weapons. States could commit to conducting regular and in-depth discussions with other nations to address tensions and prevent crisis situations.

- A Global Fund for Nuclear Disarmament

A global fund for nuclear disarmament could be created, funded by voluntary contributions from interested nations and individuals. This fund could support projects for the conversion of nuclear infrastructure, awareness initiatives, and peace education programs.

- Utilizing Technology for Verification

Technology can play a crucial role in nuclear disarmament

verification. The use of satellites, drones, and advanced monitoring systems could provide an unprecedented level of transparency and control over nuclear disarmament.

- The Non-Use Agreement of Nuclear Weapons

Nations could consider the possibility of entering into a global agreement not to use nuclear weapons under any circumstances. This could represent a significant step toward disarmament.

- Active Involvement of New Generations

New generations should be actively involved in promoting nuclear disarmament. Educational programs, hackathons, and creative competitions could be organized to encourage young people to contribute innovative ideas to disarmament.

- The Role of Art and Culture

Art and culture can play a fundamental role in promoting nuclear disarmament. Film festivals, art exhibitions, and music concerts dedicated to nuclear disarmament could engage the public emotionally and creatively.

- Regional Cooperation

World regions could develop regional cooperation initiatives for nuclear disarmament. Nearby countries could collaborate to promote the reduction of nuclear arsenals and stability in their region.

- Financial Transparency

Nuclear nations could commit to providing complete financial transparency regarding military spending related to nuclear weapons. This would allow for a clearer understanding of the size and intentions of nuclear arsenals.

- Engagement of Celebrities and Public Figures

Celebrities and public figures can play a crucial role in promoting nuclear disarmament. Their involvement in awareness campaigns and public support can amplify the voice of the movement.

- Online Mobilization

Online and social media campaigns can mobilize millions of people worldwide. Creative use of hashtags, online petitions, and viral videos can draw global attention to nuclear disarmament.

- Economic Sanctions for Non-Compliance

Nations could consider adopting economic sanctions against countries that do not comply with nuclear disarmament agreements. This could be a means to create an incentive for compliance with commitments.

Conclusion

Nuclear disarmament is a global challenge that requires innovation, commitment, and creativity. In addition to highlighting challenges and opportunities, it is important to consider original ideas to accelerate progress toward a world without nuclear weapons. With global unity and innovation, the dream of a nuclear-free world can become a concrete reality.

EPILOGUE

As we reach the conclusion of this journey through the world of nuclear, let us reflect on what we have discovered and the steps ahead. "Stop the Nuclear: Dreaming of a World Without Nuclear Energy and Atomic Bombs" has been an exploration of the many facets of nuclear, a dive into its scientific, political, and human implications.

Nuclear is a topic that shakes consciences, raises deep concerns, and challenges our sense of responsibility towards the planet and humanity itself. Throughout our journey, we have touched upon the roots of nuclear energy, examined the horrors of atomic weapons, and delved into the complex implications of nuclear policy.

We explored the origins of nuclear energy, from the discovery of the atom to its early practical applications. This journey has led us to a deeper understanding of the challenges and opportunities associated with nuclear energy.

We then confronted the catastrophic potential of nuclear weapons, recognizing their capacity to inflict widespread destruction. These weapons, despite their terrible effectiveness, also played a role in maintaining global balance during the Cold War.

The painful lessons of Fukushima and Chernobyl taught us about the long-term consequences of nuclear accidents, highlighting the importance of nuclear safety and responsible nuclear plant management.

We examined the hidden costs of nuclear energy, revealing its often underestimated environmental and health impacts. It became clear that nuclear energy involves significant trade-offs that go beyond its apparent efficiency.

The influence of the nuclear industry on politics was explored, underscoring how economic interests can shape government decisions. This raises critical questions about transparency and accountability in nuclear policies.

International organizations such as the International Atomic Energy Agency (IAEA) and the Treaty on the Non-Proliferation of Nuclear Weapons (NPT) gained relevance, playing a fundamental role in nuclear control.

The next chapter introduced us to the Global Nuclear Disarmament Initiative, highlighting the contribution of global movements for peace and nuclear disarmament. These activists and organizations push for significant changes in nuclear policies.

We then examined the efforts of nuclear powers to reduce their arsenals, recognizing the complexity and delicacy of nuclear disarmament. The reduction of nuclear weapons is a crucial step towards a safer world.

Science and technology, as we have seen, can support the cause of nuclear disarmament through creative innovations. Scientific research is a potent ally in the fight against nuclear challenges.

We also explored the role of charismatic leaders in promoting nuclear disarmament, showcasing examples of global figures who influenced global politics.

The book also shed light on peace and nuclear disarmament movements worldwide, demonstrating how ordinary people can make a difference through activism and awareness-raising.

The voices of nuclear victims were heard, giving them a space to tell their stories and provide eyewitness accounts of the devastating consequences of nuclear accidents.

We examined the role of art and culture in promoting awareness of the nuclear issue, recognizing the power of artistic expression in conveying emotions and reflections on the nuclear reality.

Furthermore, we painted a picture of a future without nuclear weapons, emphasizing the benefits of a peaceful world and our commitment to this goal.

Geopolitical tensions hindering nuclear disarmament were examined, highlighting the complexity of the context in which this struggle for peace takes place.

The importance of education on the nuclear issue was recognized, with education playing a fundamental role in promoting awareness and understanding.

We also explored the impact of non-governmental organizations (NGOs) in advocating for nuclear disarmament, emphasizing the significant role they play in shaping global policy towards a world without nuclear weapons.

The book emphasized individual responsibility in nuclear disarmament, highlighting that each of us can contribute to this cause through daily actions.

Finally, a concrete plan for a world without nuclear was offered, a roadmap that requires commitment and action.

As we close these pages, we cannot help but wonder about the future of nuclear and nuclear disarmament. We are aware of the challenges we must face, but we must also keep alive the hope for a safer and more peaceful world.

Nuclear can be a creative or destructive force, but the direction it takes depends on the choices we make. Together, we can make a difference, working towards a future where nuclear energy and atomic weapons are no longer a threat to our existence.

We hope that this book has inspired you to reflect on nuclear in new and meaningful ways, prompting you to consider the role you can play in shaping our nuclear future. May it have made you dream of a world without nuclear, but also take action to realize that dream.

Thank you for being with us on this journey. The road may be long and winding, but it is a road we must travel together. The future is in our hands, and it depends on us to make wise and responsible choices.

With hope and determination,

John Valentine

AFTERWORD

Now that we have completed this journey through the world of nuclear, it is time to reflect on what we have discovered and how we can move forward. We have explored the roots of nuclear energy and its impact on our society, confronted the threats of nuclear weapons, and examined how politics and science intertwine in this complex field.

We have seen how nuclear energy has the potential to provide a significant source of electricity but also entails significant environmental and health risks. We have examined the tragic moments of Fukushima and Chernobyl, which taught us the harshest lessons about nuclear safety and its consequences. We have explored the political implications and the role of international organizations in nuclear control.

We have also looked ahead, examining the paths that could lead us to a world without nuclear. We have discussed the contribution of science, the role of charismatic leaders, and the power of activism and art in shaping our nuclear future.

But in the end, we should ask ourselves: what can we, as individuals, do to contribute to a world without nuclear? The answer is that each of us has a role to play. We can educate ourselves and others about the nuclear issue, support organizations advocating for nuclear disarmament, and make our voices heard with our political leaders. In this complex and sometimes frightening world of nuclear, we must be informed and active citizens. We must dream of a world without nuclear, but we must also work to make that dream a reality. The road will be long and challenging, but it is a road we must travel together.

I hope that this book has inspired you to reflect on nuclear in new and meaningful ways. That it has prompted you to consider the role we can play in shaping our nuclear future. That it has made you dream of a world without nuclear but also take action to realize that dream.

Thank you for accompanying me on this journey. Nuclear can be an ally or an enemy, but its direction depends on the choices we make. Together, we can make a difference.

With hope and determination,

John Valentine.

ACKNOWLEDGEMENT

I would like to express my sincere gratitude to all those who have chosen to embark on this journey through the pages of "Stop the Nuclear: Dreaming of a World Without Nuclear Energy and Atomic Bombs." This book has been a labor of love and dedication, and your interest and support mean a great deal to me.

First and foremost, I want to thank the readers. You are the reason I wrote this book. I hope you have found these pages informative, thought-provoking, and engaging. Your curiosity and interest in nuclear and nuclear disarmament are essential in promoting a deeper understanding of this crucial issue.

Special thanks go to those who have directly contributed to this project. To those who have shared their knowledge, experiences, and perspectives, allowing me to create a more comprehensive and balanced book. To researchers, activists, experts, and everyone who has shared their time and wisdom, thank you from the bottom of my heart.

I also thank my family and friends for their constant support and encouragement. Writing a book is a commitment that requires time and dedication, and your support has been instrumental in completing this project.

Special thanks also go to the OpenAI team, who have made

the creation of this work possible through advanced artificial intelligence technology. Your commitment to innovation has opened new possibilities in writing and sharing knowledge.

Finally, I want to thank anyone working for a safer and more peaceful world. Whether they are activists, diplomats, scientists, teachers, or leaders, your commitment to nuclear disarmament is of paramount importance. Keep working for a future without nuclear weapons because it is a goal worthy of every effort.

Together, we can make a difference. Thank you again for reading "Stop the Nuclear" and for being part of this important dialogue on the nuclear issue.

With gratitude,

John Valentine.

ABOUT THE AUTHOR

John Valentine

John Valentine is an Italian author born in 1986. After completing his university studies, he decided to embark on a journey around the world to deepen his understanding of various cultures and spiritual practices. During these travels, he discovered new techniques that helped him develop his passion for writing and exploring the mysteries of the universe.

Valentine paid particular attention to themes related to energy, attraction, and chakras, learning various techniques for balancing and purifying his own energy centers. His knowledge and practice of these techniques allowed him to gain greater awareness of the importance of energy in daily life and interpersonal relationships.

Upon his return to Italy, John Valentine decided to share his knowledge and experiences with the world through writing. His first book, "The Warrior of Knowledge: Tap into the True Potential of Your Mind with Positive Thinking," became an international bestseller due to its ability to clearly and accessibly convey his understanding of energy.

Valentine continues to write and share his knowledge with the world, helping people improve their lives through the understanding and conscious use of energy. His passion for writing and ongoing research continues to inspire many people worldwide, and his commitment to spreading this knowledge remains strong and unwavering.

BOOKS BY THIS AUTHOR

The Knowledge Warrior: Tap Into The True Potential Of Your Mind With Positive Thinking

This manual tells the story of a boy and his compelling and unusual life journey. Born and raised in an era where society appears to have lost its values, living in hypocrisy and falsehood, driven by a mentality built upon false ideals, the boy starts asking questions and seeking answers through his own introspection. He looks beyond the barriers imposed by society and conventional logic. It's a capacity we can nurture, develop, refine, and pass on to enhance our relationship with ourselves, with others, and with the reality we encounter every day. We can choose to live with love and positivity, being happy and grateful for this wonderful and magnificent adventure that is life.

Prepper: Survival Enhanced: The Ultimate Guide To Prepare For Any Emergency And Live Safely And Independently

"Prepper: Survival Enhanced" is the ultimate guide for those who want to be prepared for any emergency and live with confidence and autonomy. Written by John Valentinc, an expert in the field of prepping and survival, this book offers a comprehensive and detailed approach to acquire the skills necessary to face any emergency situation with confidence and determination.
In the chaos and uncertainty of the modern world, being prepared becomes essential. "Prepper: Survival Enhanced" takes you step-

by-step through all aspects of preparation, from recognizing potential threats to creating a personal emergency plan, from getting water, foraging for food, building a shelter , lighting the fire, first aid and much more.

Valentine shares her in-depth knowledge of survival techniques, offering practical advice on obtaining and purifying drinking water, storing food long-term, hunting, fishing and gathering food in the wild. You will also learn how to build temporary and permanent shelters to protect yourself from the elements and how to make fire in different situations. Plus, you'll learn the basics of first aid and how to deal with common injuries in emergency situations.

But "Prepper: Enhanced Survival" isn't just limited to basic survival skills. Valentine also guides you in managing stress and mental well-being during critical situations, in orienteering in uninhabited areas, in self-defense and protection, in managing limited resources and in using the technologies and tools available for survival.

This book is suitable for everyone, from beginners who want to get into the world of prepping to experienced preppers who want to improve their skills. The detailed explanations, practical instructions and invaluable suggestions provided by Valentine will allow you to develop a personalized preparedness plan that fits your specific needs and gives you the peace of mind knowing you can deal with any emergency situation.

It doesn't matter if you live in the city or in the countryside, whether you are planning for a natural event, a pandemic or a man-made disaster, "Prepper: Empowered Survival" is your complete and reliable resource to successfully meet any challenge. Get ready for a safer future today and buy John Valentine's "Prepper: Enhanced Survival".

The Secret Of Attraction: The Power Of The Mind And The Art Of Creating The Desired Reality

The book "The Secret of Attraction" by the famous author John Valentine is a complete and detailed guide that leads you towards discovering the power of your mind and its ability to create the reality you desire. Based on the powerful Law of Attraction, this book offers a practical and transformative approach to transforming your life.

Through an engaging narration and a series of real examples, the author explains in a clear and accessible way the fundamental principles of the Law of Attraction. You will learn how your thoughts, emotions and actions can directly affect the energy you attract into your life. You will discover that you are the creator of your own reality and that you have the power to manifest your deepest desires.

In the book, John Valentine takes you step-by-step through the process of applying the Law of Attraction. You will learn to recognize and overcome the limiting beliefs that prevent you from achieving what you want. You will be provided with powerful visualization techniques and affirmations to program your mind with positive, focused thoughts on your goals.

But "The Secret of Attraction" doesn't just stop at theory. It also gives you practical tools to integrate the Law of Attraction into your daily life. You will discover how to cultivate gratitude, love and compassion as forces of attraction. You will learn to use intention and positive energy to align your vibration with that of the universe, thus opening the door for your dreams to manifest.

This book is not just a simple manual, but a comprehensive resource that takes you on a journey of personal transformation. You'll find practical exercises, guiding questions, and self-reflection tools to help you explore your limitless potential and overcome any challenges you may encounter along the way.

Whether you're just starting your journey into the Law of Attraction or are already an experienced practitioner, "The Secret of Attraction" offers a fresh and inspiring perspective that will help you achieve tangible results. No matter what your desire, be it love, prosperity, health, or happiness, this book provides you with the knowledge and tools you need to make your dreams a reality.

Revolutionary Mindfulness: A Revolutionary Guide To Supreme Awareness And Inner Transformation

"Revolutionary Mindfulness" by John Valentine is an extraordinary book that will take you on a journey of self-discovery, opening the door to supreme awareness and inner transformation. With a unique and innovative perspective on the practice of mindfulness, the author offers powerful tools and teachings to embrace a more authentic, joyful and meaningful life.

Through the pages of this book, you'll be guided through practical exercises and enlightening insights that will help you develop a profound awareness of the present moment. You will discover how to free yourself from the limiting mental patterns, fears and worries that often imprison us, paving the way for a fuller and more rewarding existence.

John Valentine shares his in-depth knowledge and personal experience, giving you a clear and understandable guide to applying mindfulness in your daily life. You will learn to develop an awareness that goes beyond the rational mind, embracing the wisdom of the heart and connecting with your true essence.

This book doesn't just offer simple instructions on practicing mindfulness, but takes you on a transformative journey that will change the way you think, feel and live. Through engaging stories, concrete examples and targeted exercises, John Valentine guides you towards a deep understanding of yourself and your relationships with the world around you.

"Revolutionary Mindfulness" is a work that goes beyond the limits of traditional mindfulness, exploring new horizons of awareness and self-transcendence. It will teach you to live in the present with gratitude, compassion and joy, overcoming daily challenges and paving the way for a life full of potential.

If you wish to explore the depths of your essence and live a more meaningful life, this book is an indispensable companion on your journey of personal and spiritual growth. Get ready to transform your life and discover a revolutionary awareness that will open the door to a life of authenticity, wisdom and fulfilment.

Reality Shifting Radical Transformation: Explore Secret Realities And Unlock Your Inner Power

"Reality Shifting" by John Valentine is a compelling book that will guide you through the fascinating world of reality shifting, a practice that allows you to explore alternate realities and transform your life in surprising ways.

Through a combination of in-depth theoretical knowledge, practical exercises and personal experiences, John Valentine introduces you to the secrets of reality shifting, giving you powerful tools to explore hidden dimensions and unlock your inner power.

Throughout this book, you'll discover a variety of reality shifting techniques, including visualization practices, lucid meditation, and creating alternate realities in your mind. John Valentine provides you with step-by-step instruction on how to use these techniques safely and effectively, enabling you to have profound and transformative experiences.

In addition to exploring reality shifting techniques, the book also examines the psychological, spiritual and cultural aspects associated with this practice. You will be guided through an in-depth analysis of the psychological effects of reality shifting and its possible scientific explanations. Additionally, the connections between reality shifting and spiritual beliefs will be explored, as well as the influence of reality shifting in popular culture.

A significant part of the book is also devoted to the personal experiences of reality shifting shared by the participants. Through these engaging narratives, you will be inspired and motivated to explore your reality shifting possibilities and to live a richer, more fulfilling life.

"Reality Shifting: Radical Transformation" is a comprehensive and accessible book that offers a balanced and informed approach to reality shifting. It is an invaluable resource for anyone interested in exploring new dimensions of consciousness, broadening their perspective, and accessing hidden potential.

Whether you are a curious beginner or an experienced practitioner, "Reality Shifting: Radical Transformation" will be your reliable guide into the fascinating world of reality shifting. Get ready to embark on an extraordinary journey that will change your perception of reality forever and allow you to transform your life in ways you never thought possible.

Archaeological Revelations: Uncovering The Mysteries Of The Past

The book "Archaeological Revelations: Uncovering the Mysteries of the Past" is a fascinating journey through some of the most important and fascinating archaeological discoveries of all time. The book explores some of history's most intriguing mysteries, such as the ancient Egyptian pyramids, the moai of Easter Island, Christopher Columbus' discovery of America, Mayan culture, and the lost city of Machu Picchu.

Overall, "Archaeological Revelations: Uncovering the Mysteries of the Past" is a book that explores some of history's most fascinating mysteries and offers a window into the complexity and richness of ancient cultures and societies. The book invites readers to continue exploring and uncovering the mysteries of the past through the Mysteries and New Theories book series.

Ancient Magic And Chakras: The Union Of Magical Practice

This book, "Ancient Magic and the Chakras: The Union of Magical Practice," explores the intersection between ancient magic and understanding the chakras. The text examines how the practice of magic can be used to balance and develop the chakras, creating a deeper connection with one's inner being and the world around us. Through the analysis of ancient spiritual texts, such as the Vedas and Upanishads, and the presentation of meditation and visualization techniques, the book guides the reader towards a deeper understanding of magical traditions and chakras. This text is an essential resource for witches, wizards, yoga practitioners, and anyone interested in developing a deeper connection with their spirituality and magical practice.